The Apes of Eden

Book 4

The House of Solomon

By Jon P. Gunn

iCrew
digital productions

iCrew Digital Productions

Chula Vista, California

COPYRIGHT

The Apes of Eden – The Journey Begins

By Jon P. Gunn

Published by iCrew Digital Publishing

iCrew Digital Productions

Website: icrewdigital.com Email: icrewdigital@gmail.com

First Print Edition: November 2014

ISBN 978-0-9851752-5-2

Cover Illustration by Annie Hobbs

Anniehobbsillustration.com

iCrew Digital Productions is an independent publisher of digital works. We support the efforts of authors who wish to self-publish in the digital realm

APES OF EDEN ON THE INTERNET

WEBSITE FOR THE APES OF EDEN:

apesofeden.com

FACEBOOK PAGE:

https://www.facebook.com/apesofeden

Follow @apesofeden on Twitter

The Apes of Eden

Book 4

The House of Solomon

By Jon P. Gunn

TABLE OF CONTENTS

ACKNOWLEDGMENTS

I met Jon P. Gunn through a Navy friend in 1983. He inspired me, entertained me, challenged me and then finally gave me one of the greatest gifts that I ever received, the manuscript for The Apes of Eden. He died shortly thereafter but made a solemn last request that I wait 30 years before sharing the Apes with the world. I am honoring that request by publishing The Apes of Eden as a trilogy.

iCrew Digital Productions started as a high school club at Hilltop High consisting of very talented individuals who are just now reaching the primes of their careers. iCrew was preceded by another group of students equally talented as a part of Southwest High Video Productions. iCrew continues with young, talented group of photographers at Mater Dei Catholic High School in Chula Vista. These kids are my inspiration and my family.

I would like to thank some of the many who the Apes of Eden begin their journey including Abigail Nieuwenhuis, formerly iCrew member Abby Beckner, who performed much of the editing on the Apes. iCrew Member Karin Street has provided much needed business sense and advice. Thanks to iCrew member Stacee Strickland, who pointed me to Annie Hobbs, who has created an inspiring cover that will bring in countless curious readers to the world of Literate Louie, the Scribe of the Tribe.

And my greatest thanks to the author, who continues to inspire me. Each time I read the Apes of Eden, I learn more about him and his wonderful legacy. He was a magician with words. Let the magic continue.

Rick Lakin

Publisher

iCrew Digital Productions

IAMBIC PENTAMETER

I'm kinda glad you got me in this fix.

You never let me use my magic tricks.

The Apes of Eden is written in rhymed iambic pentameter thus falling into the category called heroic couplets. Each line has 10 syllables and the pairs of lines rhyme.

While many lines were written in 10 syllables in very early Latin, it was Geoffrey Chaucer who added the meter and originated iambic pentameter in the English Language which was further immortalized in many plays and sonnets of William Shakespeare.

THE APES OF EDEN

BOOK 4

THE HOUSE OF SOLOMON

BY JON P. GUNN

PROLOGUE AND INVOCATION

AN INTRODUCTORY SEGMENT

FROM THE JOURNEY BEGINS

My present goal is briefly to describe
the mighty deeds of Eden's famous Tribe
from high antiquity to modern times
in lucid, readable Heroic Rhymes
that nearly any member of our band
with brains between his ears, can understand.

Permit me, first, a word on scribal Style:
Traditions have been fixed for quite a while;
and I shall follow them, however it
displeases readers lacking taste and wit.
I write as Eden's classic authors wrote.

Wherever possible, I try to quote
their golden iambs. Thus should every scribe
of Eden keep the records of the Tribe.
All tastes are not alike, I realize.
I nonetheless decline to compromise!
What compromise is possible? One type
of reader savors Art, the other, tripe.
There is no tepid "Middle Way" to go.
Like death, or pregnancy, it's Yes or No--
You'll either be enraptured by the terse,
majestic cadence of Heroic Verse
or else it drives you screaming up the wall.
There seems to be no neutral ground at all.
I don't expect my work to go to waste.
We have, among us, apes of cultured taste:
the Literate Elite. I write for those.
Let lowbrows read some Scribbler's dreary prose.

This book will fill some long-felt needs. For one,
our present tribal Archives weigh a ton.

The custom of inscribing all our lore
on chiseled slabs of stone, as heretofore,
has meant that history has put on weight
beyond my competence to estimate--
much less to carry. Nor have volunteers
come forth to share this load I've borne for years
through prairies, mountains, deserts, swamps and sloughs
as Eden's Tribe it's holy Quest pursues.
A sedentary race of stay-at-homes
is justified in storing weighty tomes,
but apes who have our quest-pursuing bent
with lighter reading ought to be content.

The land through which we travel now is rife
with deadly hazards to one's health and life:
explosive gas from smoking fumaroles,
and dragons darting out of clefts and holes,
and winged snakes, concealed in clouds of gray
volcano smoke, to pounce upon their prey.
It's difficult for someone to compete

unless he's quick and agile on his feet,
and not encumbered by a heavy sack
of Literary Treasures on his back.
And if the harsh, relentless truth be told,
I'm past my prime of life: I'm getting old.
A lighter load would be a boon to me.
I'm not as nimble as I used to be.
Besides, too many "documents" we keep
are uninformative--a cumbrous heap
of sherds and fragments which cannot be read;
or commentaries, made by scribes long dead,
on writings which were subsequently lost
among the swamps and deserts we have crossed;
or whopping tales the ancient used to weave,
which now not even gaffers can believe
or myths our Senior Tribesmen have themselves
devised--of goblins, pixies, spooks and elves--
which scribes with little talent of their own
as chroniclers, immortalized in stone.
We have some archives which exist in three

or four editions, none of which agree.
Least valuable of all our books are those
in cryptic tongues no living tribesman knows.
Be sensible: Do antiquaries need
archaic texts which none of us can read?
There's too much trash the Tribe of Eden owns
of which in vain we ask: What Mean These Stones?

The monsters we contend with, day by day,
have proven helpful, in a passive way.
The fuming pits abounding in this land
provide the chemicals with which I've tanned
the dragon-leather which, if bleached, makes quite
a choice material on which to write.
One modest credit more: I'm first to think
of using dragons' soot-black blood for ink.
If all goes well, I shall present the whole
of Eden's history on one small scroll,
in style and portability improved,
with errors and obscurities removed--

a text which may an inspiration be
to apes, throughout our future destiny,
instead of merely stirring up dissent
by giving quibblers grounds for argument.
In place of many clashing texts, I give
you one, coherent and definitive!
No "food for thought" is lost. I promise that.
I save the meat, although I trim the fat.

For this first time in many days, the view
is relatively peaceful--with a few
small winged serpents soaring in the night;
but not a single Dragon is in sight.
I'll seize this priceless opportunity
to finish editing my History
before another troupe of monsters choose
to interrupt me. Let's get started, Muse!

Literate Louie
Scribe of the Tribe

December, 19067

The Fourth Book of Louie

Commonly Called

The House of Solomon

INTERLUDE

That desiccated eyesore of a plain
had never felt a single drop of rain.
The sharpest grit the winds had ever borne
had rasped the colored buttes, and left them worn
to shapes of which we'd never seen the likes:
stone bridges leading nowhere; corkscrew spikes;
outlandish, knobby, double-legged peaks
like dancing drunks or mythologic freaks.

The blazing colors dazzled every eye;
and patient vulture circled in the sky--
black spots against the turquoise, watching for
the travelers who fell, to rise no more.

The sun stared flaming daggers from the sky.

By tens and scores, the Tribe began to die.
We started losing heatstroke victims first.
Another quarter of us died of thirst.
In every hole and crevice, snakes would lurk
to strike in buzzing streaks, then proudly smirk
while some unlucky tribesman grabbed his foot
and sat there swearing till he went kaput.
Then big, misshapen birds flew down to feed
upon his flesh, with gross, repulsive greed.

The glaring sun, suspended overhead,
kept gushing heat like so much molten lead.
Our partly-roasted flesh was wracked with pains.
The sun-glare hammered on our scrambled brains.
Our tongues blew up to twice their normal size.
Mirages danced before our bleary eyes.
The only things to shade us from the glare
were flocks of buzzards, circling in the air.

With feeble, veering tread, the Tribe pushed on

till, dazed and feverish, we came upon
some water--it's the last we ever saw
for centuries--collected in a draw:
a brackish puddle, thick with windblown grit.
Beside ourselves with thirst, we drank from it.
It gave us strength to stagger on ahead.

Some troublemaker warned, "We'll soon be dead!
Let's turn back now, before we've all expired!"

We cudgeled him till all of us were tired
and left his corpse behind us, in the east.
Our guardian angels circled down to feast.

Despite the parsimony of the land,
a lot of things survive in desert sand,
like gila monsters, scorpions and snakes,
and also huge, fantastic firedrakes.

The roar of brazen wings we'd scarcely heard
before the hulk, whose speed its outline blurred,
outgrabe its hunting scream above the land,
and rushed upon our stumbling, starving band
with snarling jaws that rip, and claws that snatch,
and scales the sharpest javelins couldn't scratch.
Sulphuric roars belched forth, a flaming gale.
It raised a tempest with its lashing tail.

Delerious with thirst, the Tribe attacked,
and clubbed it till its stony braincase cracked.
The monster finally died, but died so hard
the buttes and dunes for leagues around were jarred.
It was the toughest game we ever slew,
and made the toughest steaks we've yet to chew.

For lack of normal game, we grew adept
at hunting things that slithered, crawled or crept.
The Tribe survived by learning to compete
on equal terms--those snakes were good to eat.

Fresh meat and gravy also helped our thirst.

The match was fair; they often got us first.

THE CITADEL

The vultures watched the Apes of Eden plod
across the salty waste, in quest of God.
One thing we've never lacked is fortitude.
Another thousand years the Tribe pursued
its Quest, across that desolate expanse
where buzzards wheel, and vague mirages dance.

An enemy we're almost sure to find
no matter where we go, is humankind.
Some manlike species were in evidence
in all conceivable environments;
and arid haunts of desert ghost and ghoul
were no exception to the general rule.
In chapters past, I've had harsh things to say
of human tribes we've met along our way;
but are they all so vicious? That depends.

Some were, in their misguided ways, our friends.

One day, as through the desert we advanced,
our Chief's patrol of scouting tribesmen chanced
to spot a lonesome citadel atop
a lofty butte. Our leader signaled Stop.
He scanned the somber view, his aspect grim;
the castle's looks did not appeal to him.

"That fort was built by human types, all right,"
he quickly ascertained. "Here comes a fight.
I guess we 'll have to take a look around
inside, to see if Logos can be found.
Though nothing's really likely to be learned,
we can't risk leaving any stones unturned.
They won't be overjoyed to see how far
our searching goes. Those humans never are.
I'll have to send some spies in first, I fear...
I heard a sneeze. Is that a volunteer?"

A road that spiraled up the mountain led
to scowling walls and turrets overhead.
The battlements were uniformly gray
and sinister, despite the glare of day.
It was a gloomy and forbidding place,
the lair of some voracious, grisly race,
as humans tend to be.
                    Our leader sent
two volunteers as spies, while runners went
to warn the Tribe to be prepared for war,
in case this town--like most we'd searched before--
was hostile to our Purpose, and preferred
to try to thwart our questing for the Word.

While wondering which spies he should appoint
to go (at mortal risk) to case the joint,
he noticed that the Sage and Scribe were there;
and these, he thought, were apes the Tribe could spare.

"Just take a look around," he told the two.
"Stay out of useless fights, and out of view.
Don't mess things up, as spies have done before.
His clues we need, not just another war.
If God, or traces of Him, aren't in sight,
then this might be a war we needn't fight.
Don't worry; if you're slow in getting back
we'll know you guys got killed, and we'll attack."

The Chief's patrol returned to camp to wait.
The volunteers trudged off to meet their fate.

"In golden ages past," complained the Scribe,
"a poet had prestige, around this Tribe.
He didn't have to double as a spy."
He glumly kicked a stone, and watched it fly,
and added, "I'd be smarter just to quit.
I get no recognition for my wit."

"You think you're persecuted!" scoffed the Sage.

"The job I've got's a grind, in <u>any</u> age!
Our Leader never wants me to advise,
but still insists I keep on being wise.
To be a Sage you've got to use your brain,
and heavy thinking gets to be a pain!
If you don't think a Sage's job is rough,
just take my place. You 'll soon have had enough!"

The poet said, "One thing I know for sure:
the job of Chief's a real sinecure!
The only thing he does, resembling work,
is making certain no one else can shirk;
yet every time you listen to him, he's
bemoaning his responsibilities!"

The spies proceeded, tossing to and fro
their tales of self-evaluated woe,
until they reached the bottom of the butte
and paused a moment for a quick dispute:
The poet thought they ought to take the road.

His archive sack, he said, was quite a load,*

---

*The sack went with him everywhere he went,
to guard our books from theft or accident.

---

and might impede their mission's progress, if
they lugged such heavy reading up a cliff.
The thinker held another viewpoint quite:
they'd been instructed to avoid a fight;
and since the roadway wasn't very wide,
it might be hard to find a place to hide
if, in their stealthy, secretive ascent,
they met a downward-headed resident.

They made a circuit of the butte, and found
the cliff rose sheerly upward all around,
and bulged unclimbably at greater height.
The Sage agreed perhaps the Scribe was right:

the only way they'd ever get that high
was either take the corkscrew road, or fly.

While they were hiking back to reach the road,
a Flying Disc came swooping past. It slowed,
returned, and hovered near the mountainside.
It looked to be some thirty-five feet wide,
of flat and circular design. The spies
set down their clubs so they could shade their eyes.
They stood with heads a-tilt and mouths agape,
examining this strange, metallic shape.
It hovered for a minute, lightly parked
on air, then soared aloft.
The Sage remarked,
"If we get caught, y'know, our mission's botched;
and something makes me think we're being watched."

The pathway up the butte was channeled through
the solid rock, like threading on a screw.
Along the outer edge, a waist-high wall

protected careless hikers from a fall
which would have ended in the fatal shock
of stopping at the bottom, on the rock.
The road went spiraling around the hill
a half a dozen times or so, until
it led them to a gateway at the top--
and there the secret agents had to stop.
The steel portcullis stood some ten yards tall
and somewhat wider, in a concrete wall
which loomed above them, rising toward the blue
until its top was nearly out of view.
The gate was studded, like a porcupine's
integument, with wicked metal spines.

But neither of the apes was greatly cowed.
The Scribe set down his sack, and called aloud,
"Wake up in there, and open up the doors!
You're being honored by some visitors!"

The Sage came forward, and began to drub

the huge, resounding portal with his club.

The agents hadn't very long to wait.
A shutter opened, up above the gate,
and there a darkly-hooded head looked out
to see what all the racket was about.

"Let's have a little service!" yelled the Scribe.
"We're here to spy this joint--from Eden's Tribe!
In towns where folks oppose us from the start,
our custom is to tear the place apart!"

The hooded head withdrew without a word,
but shortly after that, the tribesmen heard
an awe-inspiring and portentous sound
of grinding rumbles from the rocky ground
--a noise that jolted them from toes to brains--
and then the clanking of enormous chains.

While asking what such auguries would bring

they noticed that the gate was opening.
A center line appeared, and then the gate
divided and began to separate,
until the spies could see the light of day
between the massive slabs of steely gray.
They stood and watched the slowly-growing cleft
as gates slid ponderously right and left.

The rumbling stopped. The gate was standing wide.
Our tribesmen nonchalantly strolled inside
to be confronted by a human host
who bore a marked resemblance to a ghost--
in shape and posture seeming to conform
approximately to the human norm.
He wore a hooded robe of charcoal black
that hid his frame completely like a sack,
and had a redolence of mold and must.
The only features they could see were just
his yellow eyeballs I staring like an owl
from inky murk beneath the bulky cowl.

He stood upon a platform, flat and round,
which hovered oddly just above the ground
but seemed to be as steady as a rock.

This apparition carne as quite a shock;
but Eden's tribesmen never heard of dread .
The Scribe thrust out his jaw, and firmly said,
"Our mob left Eden, centuries ago,
to rove this crumby planet to and fro
in search of Logos, God or Deity,
or any traces of Him we could see.
We'll all be here to search tomorrow, and
there's one thing you guys better understand:
We don't like getting laughed at. Furthermore,
the way we search a place is door to door
and inch by inch. When we get done, we'll go;
but one reply we don't accept is No.
We'll finish what we come for, right or wrong;
so don't act hostile, and we'll get along."

"Perhaps," the human said, "we could direct
your search--though not in ways you might expect.
We, too, have searched, though on another path,
and seen what you have not--the aftermath."

The oddest thing about the human's voice
was that his hearers seemed to have their choice
of where it carne from. They could hear the sound
from five or six directions, all around.

The Scribe rejoined, "No phony leads of yours;
our Tribe don't need advice from amateurs !
No walls or weapons ever long deterred
our gang from searching places for the Word;
and no one's ever fooled us. Some have tried
to cloud our minds, and consequently died!"

The human chuckled hollowly. "I sense
a somewhat overweening confidence
in beasts who might be wiser to respect

superiority of intellect.
You 'll quickly lose a lot of your aplomb
when you're confronted with a cobalt bomb
...However, let's assume you spoke in jest.
Your Tribe, you say, is on a holy quest.
How nice for you. We humans sympathize
with noble goals like yours. We realize
that just such random 'questing' led to our
incalculable technologic power;
so pithecans and humans, it would seem,
have shared a sort of common racial dream.
If you can curb this choler you display,
we might agree to guide you on your way.
We know the path, you see, which must be trod
by those who'd seek a realistic 'god'."

The Scribe did not reply, but looked perplexed.
The Sage took up the conversation next.
"Well, stranger," said the thinker, "if it's true,
I'd call that real neighborly of you."

“Then welcome,” said their host. “I said before
you're ill-advised to take us on in war
--our strength is not to be disparaged--yet
you needn't take that as a counterthreat.
Long, long ago, we humans had our fill
of warfare, so we bear you no ill will.
Your nosiness derives from motives pure,
and so-- (He turned) --you'll get the Guided Tour.
Just follow me. I'll show you what we've done
to end the quest your race has just begun.

“Although we've left your species far behind,
your manifest activity of mind
entitles you, at least in some degree,
to share our common Primate Destiny.
The fact that we've been able to advance
while you could not, was partly due to chance.
We found the way, while you did not. Bad luck
was all that dug the rut in which you're stuck.

This needn't be, and with the impetus
of Education you'll receive from us
(if you accept it) you'll attain the place
that Intellect confers upon a race."

The charcoal-colored buildings, left and right,
rose windowless and blind to lofty height.
The human led the way along the street.
He didn't touch the pavement with his feet,
but floated grandly on his metal raft
like milkweed seeds that ride a gentle draft.

The apes, behind him, quietly conversed
to make their plans if worst should come to worst.
The Sage observed, "This could, y'know, be a
locale for catching claustrophobia.
They sure have got a somber-colored town.
I feel conspicuous, just being brown,"

"We can't bug out; there's spying to be done,"

the Scribe replied. "Besides, it's two to one--
this jerk's no match for us. There's still no threat.
We've seen no other human beings yet.
Let's get this place all spied, and save the fuss.
Our Leader'll be real proud of us."

Their guide approached a building. As he neared,
a doorway preternaturally appeared.
The mechanism wasn't clear: the wall
dissolved from view and wasn't there at all
within an archway, maybe five feet broad
and seven tall.

The apes weren't overawed.
Completely confident that they would win
if strife arose, they calmly sauntered in.

PERFECTED MAN

"The Primate quest for Truth," their guide began
"took vastly different forms, for Ape and Man.
Your own approach--a random search around
the Earth--is scientifically unsound.
The limitations of the eye and ear
we humans have discovered interfere
with one's perceptions of the universe;
so hardly any method could be worse
than trusting senses clearly to perceive
the subject. Such directness is naive.
There are phenomena the eye and ear
when unassisted, cannot see or hear
and which--although they readily affect
our instruments--our senses can't detect.

It will, for instance, likely be a shock
to learn that what you think is solid rock
consists of little more than empty space
through which neutrino particles will trace
as straight a course as through the empty air.
If data aren't interpreted with care,
you're sure to draw erroneous conclusions
based on simple optical illusions.
Bear in mind, the meaning and the end
of Science is completely to transcend
the limitations which your own physique
imposes on the data that you seek.
The Scientific Way to search for gods
is just like classifying arthropods
or mice, or elephants, or apes, or snarks,
or leptons, J-psis, tachyons and quarks.
Hypotheses are easiest to test
which can be mathematically expressed.
One finds the patterns in the data known
to see what Basic Principle is shown,

and then extrapolates it to explain
whatever gaps in knowledge may remain.
Though observation is the place to start,
until Evaluation plays its part
no data which are merely sensory
can build a model of Reality--
a model only, true, to that extent
that it predicts controlled Experiment.
Unless you show that all predictions jibe
with what unbiased instruments describe,
your theory's open to a valid doubt.
All failures must be firmly weeded out.
You apes have let your sheer tenacity
engender cortical opacity.
A Scientist is ready to revise
the instant new discrepancies arise.
Our first progenitors were much like you,
not just in shape and size, but point of view.
We too, in ancient times, went far astray,
before we found the Scientific Way.

"But more of that anon. Permit me first
to show that our approach is reimbursed
with tangible, applicable effect,
which proves our methodology correct."

The apes were led through hollow-echoed walls,
where eerie light exuded from the walls,
until the human took them through a door.
They found a vault, with implements of war
displayed in cases, or arrayed on racks,
and labeled "spear" and "mace" and "battleaxe."
He led them past displays of bows and slings
and arrows hafts and arbalests and things.
They saw an iron suit, with lance and shield,
and other gear that human warriors wield;
and then the relics of more recent wars,
with telescopic sights, and rifled bores.

"In <u>very</u> ancient times," their guide resumed,

“it seemed our race would be forever doomed
to unremitting and unequal strife
with specially-adapted forms of life
--who nearly wiped us out, to tell the truth--
the cave bear, mastodon, and sabertooth.
So naturally, in rowdy days of yore,
our prime concern was instruments of war.”

“The same with Eden's Tribe,” the Scribe replied.
“In olden times, a lot of critters tried
us apes for dinner; but they quickly learned
the compliment was apt to be returned.
We started making weapons out of stone,
with hefty helves of wood or bison bone.
We made us hatchets, spears, and all the rest.
Of course the good old bludgeon's still the best.
A spear's okay; but when the war gets hot,
it's hard to see if someone's speared or not.
But with a club, no matter how you’re rushed,
there's no mistaking when a skull's been crushed!

A head that's busted has a different feel
than one that isn't. That's a club's appeal."

"A most inventive race," the human laughed,
"to think of mounting spearheads on a shaft!
But there was so much more you might have done,
and even more of your encounters won,
if you'd renounced your atavistic quirk
of liking closeness to your heavy work.
We humans had a healthier respect
for crafty subtleties of intellect,
and worked some rather urgent problems out
by methods you'd consider roundabout,
but which, in actuality, were far
more efficacious than brute muscles are.
You apes on brawn and ignorance relied,
and stopped advancing, wholly satisfied.
We might have made the same mistake, and spent
all time in static nondevelopment;
but our successes, unlike yours, have led

to more inventions, so we forged ahead.

"It wasn't any easy thing to do.
We started out with even less than you.
A man, deprived of artifacts, would stand
but meager chance of living off the land.
Incapable of camouflage or flight
and all but helpless in an open fight,
the only thing that got us off the hook
was subtlety, so that's the course we took.
Besetting enemies with traps we laid
--the logfall, covered pit, and ambuscade--
we compensated with the use of these
for Natural Man's innate deficiencies.
Our racial limitations were, for us,
a constant source of mental stimulus.
For us, a problem's there so it can be
resolved by Human Ingenuity.
Our Adaptation is, and shall remain,
the awesome power of the human brain."

As apes and guide continued to advance,
the thinker gave the Scribe a sidelong glance
and raised one eyebrow, to convey his doubt
of what their host was blathering about.
The Scribe made no remark their host could hear,
but waved a finger vaguely at his ear
and crossed his eyeballs, with a grin inane,
implying that he thought their host insane.

At last the little group arrived at where
an armor-clad conveyance, huge and square,
was squatting on its iron-cleated treads.
Its phallic cannon loomed above their heads.

"We had some crude and inefficient ways
of fighting wars, in prehistoric days,"
the guide continued, "sending in a flood
of human ·fighters--muscle, bone and blood--
to grapple physically with enemies.

In more sophisticated eras, these
contrivances were standardized in war,
to cut the casualties we so deplore.
Its guns could pulverize the stoutest walls.
The caterpillar treads on which it crawls
are armed at intervals with metal lugs
it used for squashing hostile troops like bugs.
Machines replaced the human fighting force.

"The change was rather gradual, of course;
and this invention was a passing stage:
It held a crew. But in a later age
we gave machines more flexibility
by using microwave telemetry.
The 'drivers,' in a distant, bombproof hole,
could operate it by remote control.
The space once used for men to ride inside
was now with better weaponry supplied
and larger stocks of cannon shells and fuel
to give it more endurance in a duel.

When such machines in battle were deployed
they couldn't be too readily destroyed;
and if they were, their crews of living men
took charge of fresh machines, to slay again.

"Still later, we devised the robot brain,
which never slept till every foe was slain,
and which reacted with the speed of light
when enemy equipment came in sight.
Whenever we were menaced by a foe
we turned the robots on, and let them go.
The only drawback was, when they were used,
their high-strung circuits sometimes got confused,
mistaking for a fort some harmless town,
and blasting thousands of civilians down.
But carping moralists were satisfied
if most disasters struck the other side;
for with a war to win, no time was spent
on mawkish and enfeebling sentiment."

"That thing won't work on _us_," the poet said.
"We'd toss a couple boulders in the tread
and stop it in its tracks; then tie that gun
in knots, to keep from losing anyone
while we were ripping it apart, to choose
the chunks of metal someone wants to use.
It might be real useful, even so
against some less resourceful breed of foe."

"It's problematical," the human said,
"how close you'd get before you're full of lead.
The smaller guns with which this tank abounds
have firing speeds of many hundred rounds
a minute. It's equipped with infrared
perceptors, so approaching foes are dead
as soon as they're within its line of sight.
They're not concealed by fog, nor dark of night.

"And here's another thing we used to use
in warfare. Pull the lever, and it spews

a deadly stream of flaming kerosine,

destructive both to soldier and machine.

It doesn't matter now. This thing's retired.

The violence our progenitors admired

arose from Instinct. Now that human brains

have been perfected, no such urge remains.

There's more to see, however. Shall we go?"

The human led the tribesmen down the row

until they found another strange machine

the likes of which few apes had ever seen.

It had a hollow nozzle from its tank,

from which a wisp of gasses hissed and stank.

The poet wiggled levers, while the Sage

took special interest in the pressure gauge.

"It's used," the guide explained, "in types of war

we hardly can imagine, any more. It took a rather hardy foe to last

a battle out, when he'd been mustard-gassed.

Less deadly fumes were used by our police
to help suppress revolts, in times of peace;
so concepts first evolved for use in war
made governments more stable than before.
This proves we humans weren't preoccupied
with warfare. New techniques that war supplied
were later used in less destructive ways.
Each scrap of data, ultimately, pays.
Without a military missile race
we'd slower been, in reaching Outer Space.
You'll find this out yourselves when you begin
the studies we intend to start you in.

"Now let me show you how we'd fight a war
if Man were ever threatened any more.
You 'll find it interesting, I believe."

The human pointed, with his dangling sleeve,
where, just ahead, a doorway stood ajar
and said, "That's where the master consoles are.

Let's have a look--but let me caution you:

Don’t handle things. Disaster may ensue.”

HIS ARMAMENTS

Inside the open door, the tourists found
a room, some twenty yards across, and round.
It held a redolence of patient doom.
A metal panel, all around the room,
was punctuated every yard of so
with round displays of darker greenish glow
lit up with brightly-glowing starlike blobs.
Below each screen were pilot lamps and knobs.

Upon a central table lay a screen
two yards across. It glowed a baneful green.
Around its circular perimeter
a radius revolved. The tribesmen were
intrigued to see, each time the trace returned
the same small flecks of luminescence burned,
then faded out, but never quite dissolved,
and reignited as the trace revolved.

“Equipment,” said the human, “such as this
would be a sneak-attacker's nemesis.
When these were first invented, long ago,
just denser masses caused the creen to glow;
but now that we're the only ones who use
metallic war machinery, we choose
to modify our radar sets to fit
our current needs. The waves your brains emit
are sensed, resolved, and flashed upon the screens.
Potential foes are flesh, and not machines.
No creature with at least a pea-sized brain
could near this fort in privacy remain.
All living things will register, who get
within the range for which this unit's set.
Right now some twenty miles is all it scans,
to monitor your band of pithecans.
This cluster, here, immobile in the east,
shows where your Tribe is camped. Each spot's a beast
with brain the size of yours. When these appeared

we watched them till a small detachment neared--
about a dozen, roaming in advance,
quite obviously for reconnaisance.
When two small spots (the two of you) detached
from them, and came toward here, we soon despatched
a plane to reconnoiter. Hanging low
above you, it appraised you as a foe.
Had you been armed with anything but sticks,
you might have found yourselves in quite a fix;
for when the robopilot senses guns,
it promptly dumps a thousand megatons,
eliminating enemies before
they know what hit them. That's 'preventive war.'
Two unarmed apes, of course, were not a threat;
we saw no reason for excitement yet.
A large contingent wouldn't have survived.
Perhaps you saw that craft, as you arrived?
...Apparently you did." (He'd seen the apes
regard each other with astounded gapes.)

The human turned a knob, and as he did
the light-green traces toward the center slid,
while others, slipping past the rim, converged.
The former images condensed and merged.

"Before you now," the human said, "appear
the populations of the hemisphere.
On this, the lens upon the ceiling sees
potential enemies' activities."

(The apes looked up, and were nonplused to spy
a giant pupilless and lidless eye
that scrutinized the radar screen below.
Both tribesmen backed away a pace or so.

"Your mental waves hold subtleties unseen
by us, who merely watch them on the screen,
but which transducers sense, and thereby know
which flecks are neutral, which may be a foe.
They know exactly what the target craves

by analyzing his cerebral waves.
If ever we are threatened by a foe
computers will immediately know
his purpose, capabilities and plans,
and make short shrift of enemies of Man's.
The siloes here concealed would spew a swarm
of missiles in an all-consuming storm.
No hiding place exists, in all the world,
to which atomic warheads can't be hurled;
and if some cave of refuge could be found,
we'd send machines that tunnel through the ground,
with cybernetic brains to guide their way,
to ferret out and massacre their prey,

"If these exhibits may be summarized,
they show that you'd be rather ill-advised
to pit mere animal sagacity
against our human brain capacity.
Some beasts presume to scorn a man's physique
because his thews are relatively weak.

That's only superficially the case.
We've vanquished every anti-human race.
Mankind alone, of everything alive,
accomplished more than merely to survive.
For Man alone, our own inferior
endowments made our race superior.
Ballistic weapons do a lot more harm
at longer range, than does a stronger arm.
At first the spear and javelin were the best
we had; then bow and arrow, arbalest;
ballistic engine; cannon; and at last
the radiation-spewing rocket blast,
and laser cannon, whose resistless beam
can slice up mountains, turn a sea to steam,
or scorch a continental landmass brown
and mow an army of a billion down."

"Great Guns!" the Scribe exclaimed. "What kind of foes attacked you, when you needed things like those?"

The human snickered. "It was quite a foe
we had to fight. You won't believe me, though.
The worst of enemies, once war began
to be our principle concern, was Man.
We seldom used such weaponry against
subhuman species. All of them were fenced
on game preserves, or tamed and brought to terms
(except for roaches, rats, and certain germs).
What feral beasts were left, we'd hunt for fun
with nothing more destructive than a gun.
Explosive charges drove our whale harpoons,
and bombs dispersed incipient typhoons;
but these were incidental. Only we
required sophisticated weaponry.

"Yet, paradoxically, when once begun
we found that such a war could not be won.
We only tried it once; and that's when Earth
as real estate, attained its present worth.
The human race became so dangerous

that even we were terrified of us.
We went to unbelievable expense
to arm ourselves for humankind's defense
against that dedicated enemy
--who was none other than Humanity.
Our finest minds and latest expertise
were all devoted to the Cause of Peace,
till all exploded, in a holocaust
of self defense which all defenders lost.
We then found out it's simpler to erase
the food supply and economic base
of rival populations, than rebuild
the ecologic systems we have killed.
We still have sentimental souls today
who wish we'd found that out some other way."

"If you're as fierce as that," the Sage put in,
"then here's a thing that I've been wonderin':
With what you've got, your power's absolute.
How come you're parked on this secluded butte

instead of out there conquering the Earth?"

The human chuckled dryly. "What's it worth
to rule the world? Our race has long outgrown
the joy of conquest--if we're left alone.
With power to deliver instant death
to everything that draws a living breath
in any part of Earth, or all at once,
who wants more power, other than a dunce?
What tribute could subhuman races pay
that we could use? With all the fine array
of fabrication systems in this fort,
the world has nothing for us to import.
There was a time in history, of course,
when we enslaved the ox, and goat, and horse,
and others who their labor could provide;
but labor now is by machines supplied.
With power hardly less than that of gods,
why bother ruling over common clods?
Who needs a slave, to serve on bended knee,

what fabricators furnish instantly?
(You'll see those, incidentally. They're found
about a kilometer underground,
where there are minerals on which to feed,
plus geothermal energy they need.)
With no desire for foreign conquest, or
the fear of conquest, we've outmoded war.
Our weapon system's lain unused more years
than you could count. No worthy foe appears.
We humans rule the Earth, without dispute.
Our power, as you said, is absolute:
but now that we're the planet's ruling race
we can't be bothered governing the place."

The poet gave a slow and thoughtful nod.
"That's quite a way," he said, "to pack a rod.
You humans sure invent a lot of stuff.
We'd like to see some more. Lead on, MacDuff."

They started through a corridor once more,

and hadn't gone a hundred feet, before
the human stopped, and waved the apes ahead.
"Go on; I'll follow right behind," he said.

The apes went on--a bit suspiciously.
Although this hadn't seemed maliciously
intended, both the apes distrusted it.

They stopped, on coming to an open pit
some ten feet wide, that reached from wall to wall,
obstructing further progress down the hall.
From deep inside the hole there rose the sound
of rhythmic thumps and rumbles, underground.

Our emissaries made one careless botch:
Each thought the other ape was keeping watch.
They leaned across the overhanging ledge
that ringed the tunnel, all around the edge,
to see what sights the blackness might afford--
until the human pushed them overboard.

Is this the end, for our intrepid apes,
or will they somehow manage their escapes?
A roaring updraft past their eardrums flowed.
Don't miss the next exciting episode!

His Livelihood

While plunging down the elevator shaft
they found themselves supported by the draft.
The windy howling gradually increased.
Their speed of fall slowed down, then nearly ceased.
They landed softly, half a mile below,
in caverns lighted by a pallid glow,
where all the apes could see for scenery
was rumbling hulks of strange machinery.
The spies stood up, and smoothed their ruffled fur,
and tried to pierce the ghostly greenish blur
to watch the bedlam of equipment moil.
The air of ozone smelled, and grease and oil.

The human made his exit from the shaft
a half a minute after them, and laughed.
"I'll bet, I had you thinking you'd be dead
when you connected with the floor," he said.

"I couldn't kill you that way if I tried.
The brakes are automatically applied
in case a human being in the shaft,
develops some malfunction in his raft.
Of course that never happens. By our arts
the raft's designed to have no moving parts,
which means there isn't any wear at all.
It can't wear out, and thus can never fall.
It's been a hundred thousand years, about,
since anybody's tried that airbrake out,
and so I thought I'd run it through the test.
I hope you're not offended by my jest."

"We're not above a little harmless fun,"
the Scribe replied, "unless it's overdone.
So take it easy, 'cause you never know
just when a person's temper's apt to blow."

The tribesmen's eyes adjusted to the gloom.
They heard their eerie tourist-guide resume:

"We humans made the absolute control
of matter, space and time our long range goal.
The task was forced upon us, in a way,
in quite an early prehistoric day,
when our progenitors could scarce contrive
to meet the competition and survive.
In order to compete on equal terms
with giant felidae and pachyderms
and other natural foes, we had to train
ourselves to use the power of the brain.
Though pit against the fiercest savage brutes,
our subtler talents yielded richer fruits;
and even in our racial origins
we clothed our nakedness in tiger skins

"Perhaps you apes have heard it said that Man
was spurred by Need alone, when he began
his long and still-continued tendency
on artifacts to place dependency.
Our clothing, as the ignorant declare,

is only worn because we've little hair

That's not quite true--or only lately so.

We did have fur, five million years ago;

but since the weather rarely stayed the same

for three successive days warm fur became

less help than cumbrance. Home-grown fur could not

be shed, like clothing, when the weather's hot;

nor was it always adequately thick

when chill December pierced us to the quick.

Eventually, the clothes we learned to wear

improved upon--and then replaced--our hair.

For pilose skin we have no further need;

our follicles have therefore atrophied.

"And something very similar befell

when hunters found they couldn't as well

as game they sought, who quickly fled in fear

before the nimrod knew his game was near.

Our first expedient was to disguise

the human hunter from his quarry's eyes.

An archer, wearing skins of buffalo,
could slyly creep in close to ply his bow.
But that was just an early, makeshift phase.
In later eras, we discovered ways
by which our sight was actually improved,
and thus our disadvantage was removed--
resulting in the further atrophy
of human visual acuity.
A modem man, to see at all, depends
- upon an artificial crystal lens.

"There's one essential point to emphasize:
such artifacts improve on natural eyes.
Out-spying eagles seemed beyond all hope
till humankind devised the telescope.
'Degeneration, when one seeks recourse
to artifacts, is thereby reinforced;
but every loss thus suffered was repaid
at least a dozen-fold by gains we made
in manifested progress toward our goals.

Our vision--hardly better than a mole's--
is now a disadvantage in reverse:
observatories scan the universe!
Although in ancient times the human race
among all beasts was slowest in the chase,
this was a blessing in disguise: a spur
to make our locomotion speedier.
We tamed the hardy horse, and taught him how
to carry us--as well as draw the plow
and haul our baggage on an A-frame sled.
Still later, we invented carts instead.
The concept was suggested when we found
that axled discs would roll across the ground
much less abrasively than would a load
that drags and scrapes along a rocky road.
We next devised the self-propelled machine:
The piston engine, burning gasoline,
replaced the horse, with vastly greater speed.

"We then attacked a deeper human need:

For ages, humankind had watched the sky,
admired the soaring hawk, and longed to fly.
That natural barrier was broken, soon,
and men ascended in a fire-balloon.
A list of ever-greater triumphs, wrought
by each successive human aeronaut,
humiliated every bird that flew.
But technologic progress wasn't through.
Each flier broke all prior records set,
and all of those were broken by the Jet.
Then spaceships landed on the moon and Mars,
surpassed the speed of light, and reached the stars.
At last, techniques of our ingenious race
enabled us to flash through cosmic space
more handily than any astronaut
by teleporting at the speed of thought.
Remind me, if we've time, to
the operation of the Spatial Gate--
a box of measured volume which exchanges
space for other space, at any range

within about a billion trillion miles
according to the setting of its dials.
We're thus to any destination flown,
provided its coordinates are known.
The Spatial Gate's no longer used; just stored,
since every place of interest's been explored.
For ordinary purposes, of course,
the last and best replacement for the horse
--this raft I ride--makes travel much more fleet
and effortless than any creature's feet.
By bloating space, it forms a cavity
which floats upon a field of gravity.
Computerized, it knows exactly where
I wish to go, and smoothly floats me there.

"So need, you see, was just the start of our
unending search for technologic power.
We're challenged less by any natural foe
than by an all-inclusive Need to Know.
You well may wonder how we here contrive

to live off land where nothing is alive,
and how we've set ourselves completely
free from local climate and ecology.
You see before you now the present base
of raw subsistence for the human race."

He opened up a cabinet standing there,
and passed out hard hats for the apes to wear,
before conducting them along the aisle
between machines that seemed to stretch a mile
ahead of them. Huge pistons rose and fell,
and bubbling vats exhaled a rancid smell.

The human waved his sleeve at vats in rows,
connected by a maze of pipe and hose.
"In early times, like any other beast,
we gathered what we could. As skills increased
we rose above the savage brute's estate
in which the moods of nature ruled our fate.
Where needed foodstuffs were not apt to grow

we planted seeds, devised the spade and hoe,
and substituted agronomic toil
for hunters' luck. But cultivated soil
could not be counted on to fill our needs
except by constant drudgery. Some breeds
of ungulates were harnessed and employed
to pull our plows. Whenever we enjoyed
a little time we weren't obliged to spend
for tasks on which our livelihoods depend,
we spent it scheming to improve the yields
of livestock herds and cultivated fields.
The bones and muscles of the beasts replaced
our own, reducing time we had to waste
to earn our bread; so 'leisure work' refined
the keen precision of the human mind.

"But there were many problems left unsolved
in which we shortly found ourselves involved--
solutions to them oftentimes result
in further problems, still more difficult.

We couldn't bank on rainfall to supply
our water, lacking which our crops would die.
So first we learned to bore through dirt and rock
to reach the water table's moisture stock,
and then to seed the clouds, and thus obtain
our water needs from artificial rain.
In coastal zones, we learned to purify
the brine, in inexhaustible supply.
The seas' resources and the atmosphere's
supplied our needs for many thousand years.
But finally, when the Earth began to die,
we found that every source was running dry.
Not even pipelines from the polar caps
were wholly free from intermittent lapse.
With natural water sources all destroyed
more drastic measures had to be employed.
We met this challenge, and renewed our stocks
of water by extracting it from rocks."

He pointed out a thundering machine

whose piston, dropping from above, between
four lubricated, smoking guiderails crashed
upon a boulder, which to rubble smashed.
A shovel, held by metal arms, beside
an open furnace door, scooped rocks inside.
More arms another man-sized boulder threw
upon the stage. The piston rose anew
in preparation for a knockout punch
upon the frightened, cringing boulder. Crunch!

"Our water is entirely now supplied
by technologic means," explained the guide.
"Dihydrogen monoxide's taken from
the hydrous sulphate formed when calcium
is left behind as dry precipitate
as ancient, dying seas evaporate.
The formula, in case you want to know,
is $CaSO_4 + 2H_2O$.
Some other salts we use are carnallite,
and, when we come across it, kieserite.

The residues are not a loss for us.
We save the sulphur, iron, phosphorus,
and calcium, and sometimes nitrogen--
with no unseemly drudgery for men.
I hope you noticed, this is done without
a human being anywhere about

"Our present water needs are much reduced,
since food is from the elements produced.
no evaporation losses, thanks
to endless-cycle hydroponic tanks."

A monstrous cauldron, hulking by the aisle
emitted noxious vapors, rank and vile.
It stood some twenty hands, and twice that wide.
The apes climbed ladders for a look inside.
"This vat's where proteinaceous yeasts are grown,"
the guide went on, "from inorganic stone.
A blend of dead reagents yields the raw
materials on which the fungi draw

to synthesize the chemicals of life.
They toil without the once-disruptive strife
of Management with Labor, or the need
to pay a Union Wage. They merely feed
--like lichens--on the crushed and slurried stone,
and gladly work their fingers to the bone.
We help them out by adding algae, to
add flavor to the somewhat tasteless brew
the fungi have to live on. Otherwise
one cannot make them photosynthesize.
We're generous enough to pay the bill
for light to drive the algae's chlororphyll.

"Our synthesis techniques are so refined
that food production needs no human mind
to supervise, yet turns out such supply
that hardly anyone has time to try
the whole selection. We have now replaced
the toil on which subsistence once was based
and freed out time for worthier pursuits,

appropriate to Human Attributes.

"Since apes resemble men, in certain ways,
you, too, will find objective science pays.
Technology, which lets your race pursue
the Mental Life, is open now to you.
The rudiments of Science we'll provide,
and some equipment--if you so decide.
You'll do this voluntarily, of course.
It's nothing we could make you do by force;
nor have we any wish to use duress
to elevate you.. It will be largesse--
a gift from men to apes.
"Now let's go see
some curiosities of history
--devices which we now no longer use--
which illustrate how humankind pursues
the same exalted goals your tribal quest
and sateless Thirst for Knowledge manifest."

His Corporeal Being

They reascended through the windy shaft,
up-wafted by the now-redoubled draft
which bounced them jarringly from wall to wall,
until it dumped them in the main floor hall.
Their guide was chuckling with annoying glee,
unruffled by the wind's ferocity.
The windblown apes resolved to ask for rafts
next time this tour included upright shafts.
Each tribesman also made a mental note
to recommend a war, if asked to vote
Suppressing anger and the Urge to Kill
required an application of the will.
Their present task, however, was to spy;
and that they would accomplish, do or die.

The corridor ahead was poorly lit
by faintly glowing walls; but once past it

they were conducted through a lighted dome
with crisscrossed aisles, where visitors could roam
among exhibits of the humans' past.
A few displays were open; most were glassed.
Though all of them were covered with the dust
of ages, they were oddly free of rust,
with all of those the visitors could see
turned on, and operating perfectly.
A pair of sacks expanded and convulsed.
Near by, a fist-sized purple globule pulsed.
The tubes connected to the pumping blob
absorbed its output with a rhythmic throb.
An object like a giant angleworm
was seen beneath its glass to writhe and squirm
but made no forward progress.

Both the apes

looked puzzled by these mock-organic shapes.
Their guide explained: "A harder war to wage
was that against infirmities of age
and illness. In our early centuries

the human race was subject to disease.
The quest for perfect health was one we stressed,
for we had enemies who could infest
our bodies: everything from ten-foot worms
to viruses, minute compared to germs.
To kill such pests was feasible enough,
but patching up our tattered flesh was tough.
Among the early, stopgap cures we used
were ones by which the patients weren't amused.
Medicaments which nearly slew the host
were used because they harmed his guests the most.
We weren't content with natural herbs and weeds
to fill our pressing pharmaceutic needs,
so medical technology progressed.
We solved these problems as we had the rest,
devising new, disease-resistant parts:
mechanical intestines, lungs and hearts.
Our highly vulnerable human health
thereby became the source of quite a wealth
of biologic knowledge, and techniques

to make repairs on faltering physiques.
We quickly came to understand the cause
--and hence the cures--for metabolic flaws.
Again, the scientific knowledge won
had from our native weaknesses begun.

"When every organ subject to disease
could be replaced with plastic parts with ease,
and every threat to health had been removed,
We didn't rest, but steadily improved.
our medical technology until
there wasn't any need we couldn't fill.
We had an adequate replacement part
for everything from genitals to heart;
and when the inventory was complete,
we learned to streamline, simplify, delete.

"The heart's a luxury that one can dump
when it's replaced by any other pump.
It was, therefore, among the first to be

replaced in function inorganically.
The lung is basically a membrane where
the blood exchanges gases with the air,
becoming quite redundant if one knows
a substitute for functions such as those.
Such chemicals as oxygen, in fact
can be absorbed through one's digestive tract.
Instead of breathing air, one simply fills
the blood with oxygen by gulping pill."

He paused before a slimy purple mass
that lay inertly underneath its glass.

"The basic purposes a liver serves
he said, "are storing glycogen reserves,
and keeping glucose on an even keel.
This tends to falter, just before a meal,
particularly when its owner's faced
with short supplies of food. We've now replaced
our early forebears' chancy food supply

with nutrients on which we can rely.
So why install a plastic liver, when
it's simpler to supply the glycogen?
The liver has some other functions: that
of storing and metabolating fat,
detoxifying blood, secreting bile
(to sour one's mood, and make one's temper vile),
etcetera. With plenty here to eat,
the liver's storage function's obsolete;
and this machine--" (He showed the apes a tank
whose ducts were oozing liquid, red and dank)
"--can manufacture artificial blood
of grade-A quality, in endless flood.
Or else one sets this dial properly
for grade- B blood, grade-O, or grade AB.
You set this Rhesus-factor switch to give
your choice of positive or negative.
This glucose-level knob, turned low or higher,
gives all the pep and vim you may desire.
You'll notice knobs and valves on every side,

so all the blood components are supplied.
This prototype, I'm sure you realize,
was later drastically reduced in size
until we had a model to install
whose weight the patient wouldn't feel at all.
Replaced by this, the liver carne to be
a mere organic superfluity.
Since blood cells can in test tubes be produced,
the need for marrow tissue is reduced;
and blood cells are themselves redundant, when
blood surrogate is charged with oxygen
that's carried to the tissue cells direct
in any concentration we select.
This means the alveoli of the lung,
as well as bones and liver, were among
the organs which a surgeon needn't waste
his time transplanting. They had been replaced
in one ingenious stroke, as you can see
by modern medical technology.
What nature had with several organs done,

we managed more efficiently with one"

"I guess, " the thinker said, "that's why we see 'em
in this antiquarian museum?"

"Yes," the guide replied. "The human heart
was not as good as its replacement part;
which also, in the last analysis,
was quite a bit inferior to this
--the HemoSynthesizer. This alone
replaces everything from lung to bone.
As such replacements are replaced themselves
they're just retired to these museum shelves."

The Scribe, by now, was looking rather wan.
He suddenly suggested, "Let's move on.
I never minded seeing bloody game;
but somehow that invention ain't the same."

The last display along the row of glassed

exhibits that the tourists wandered past
was like a giant walnut on a stem
all molded plastic. It appeared to them
about the size and shape a brain should be.

They eyed the walnut apprehensively.
It was connected, by a copper nerve,
to clockwork gears and wheels, which seemed to serve
as mechanisms for the plastic eye
that turned to watch them, as they sidled by.

His Education

The human led the tribesmen down the hall
and opened up another outside wall.
They went outside, then down the dingy street.
The tribesmen plodded on their natural feet
while their informant floated on ·the disc
of weight-defying tin, above the risk
of stubbing toes upon the path he traced,
since toes had evidently been replaced.

This time they hadn't very far to go
to see the latest types of curio.
The human opened still another wall
to show them still another vaulted hall.
They here beheld a different kind of sights:

sheet metal casings, decked with winking lights.
The tribesmen gazed around the room, agog,
while their informant pressed his monologue:

"The purpose " he resumed, "our body serves
is holding up the ganglion of nerves
we call the Brain, which is in turn designed
to carry on the functions of the Mind.
And so, however physical we seem,
the Intellect--for humans--is supreme."

The tourists followed their loquatious guide
along the aisle, as warily they eyed
the strange inventions of the human race.
The tribesmen lagged behind the human's pace.
Proximity to that uncanny shape
made icy chills go crawling up one's nape.

"In duplicating anything, one starts
by analyzing its component parts.

And once it's been completely analyzed
the working parts, of which it is comprised,
can then be duplicated, one by one,
so everything it does is better done
by faster, more efficient mechanisms.
Neurons have become anachronisms.
Here, you'll witness what our race designed
to strengthen and extend the human mind."

He brought them to a section of the room
where twinkling consoles brightened up the gloom.
Their scintillating panels (interlinked
by shielded cables) twinkled, flashed and blinked.

"When we had vanquished all our other foes,
pursuit of Knowledge was the course we chose
to siphon off the mental energy
we'd needed for survival, formerly.
We won; and, as in every war we'd won,
our victory was somewhat overdone.

We weren't content with merely knowing more
than anyone had ever known before.
The more we learned, the more we grew obsessed
with all we'd still to learn, and sought the rest.
In early times, we paid but little heed
to where 'exploding knowledge' had to lead,
till data started to accumulate
at what became an exponential rate.
We found it necessary to divide
it in to 'fields,' and let each man decide
his 'specialty'--e.g., a gnat's left ear--
to which he could devote his whole career.
Our science grew so subdivided that
not even experts knew the Total Gnat.

"The methods earlier designed to cope
with this proliferation, held no hope
that scientific knowledge could remain
the undisputed province of the Brain.
It quickly reached the point where very few

--if any--scientists could say they knew
as much as one ten-thousandth of the vast
amount of information we'd amassed.
There were no 'fully-educated' men.
The problem reached a real crisis, when
we found each 'field' was growing faster than
it could be taught or learned by mortal man.
There seemed to be no <u>terminus ad quem</u>.
New knowledge was a flood we couldn't stem
if we had wanted to. It never ceased
to grow, but exponentially increased.
That means the <u>rate</u> at which the <u>rate of growth</u>
was growing, also grew.

"I see you both
look rather doubtful--maybe just confused
by certain unfamiliar terms I've used.
I think a multiple derivative
would be the only concrete way to 'give
a dearer picture. Be content to know

that apes will have a rocky row to hoe
in utilizing knowledge we provide,
unless you also let our methods guide
your education, as your Tribe retrains
itself to give up brawn, and use your brains.

"Explosive progress hadn't reached its peak,
and yet was faster than a man could speak,
and therefore faster than a pedagogue
could teach. Nor could tech-writers catalogue
much more than sketches of it in a book.
And then, for scientists to <u>find</u> it, took
more weeks of searching overcrowded shelves
than just to rediscover it themselves.
This led to duplications by the score
of research work completed years before.
This illustrates how problems issue from
solutions to the ones we've overcome:
It had become essential to invent
a memorist-and-teacher supplement.

"The crisis passed, when DataFetch machines replaced librarians and books . This means that information, in computer code, directly and immediately flowed from laboratories into data-banks."

He pointed out a row of bulky tanks. "Those casings house the 'memory' which stores new information, in magnetic cores. Each cubic-millimeter chip contains more data than a thousand living brains. Just tell the Input what you want to know, and watch the waterfall of knowledge flow.

"One limitation was, of course, the speed at which a human scientist could read. Our next objective, therefore, was to gain direct connection to the living brain. A technical description of the means

by which we linked our Minds to these machines
would be above your heads. Suffice to say
encephalography prepared the way.
Electric pulses, which the brain emits,
could be encoded in dyadic BITs.
The only real difficulty found
was how to turn the data flow around
and force computer-generated waves
back through our skulls. The brain itself behaves
adaptively. It changes symbol mode,
and learns to think in pulsed, dyadic code.

"With that one solved, all knowledge could be taught
at speeds which far exceeded that of thought,
so learning time was drastically reduced.
A college education could be sluiced
from this machine to you, in less than ten
quintillionths of a microsecond, when
you plug these two transducers in your ears--
more knowledge than you'd gain in twenty years

of constant reading."

"Great," the Sage opined;

"but there's a question, lurking in my mind.

If this invention, like you say, contains

as much as quite a stack of human brains,

you'll get so smart your skull is apt to burst,

unless it drives you raving crazy first."

"No danger there," the human laughed. "A fact,

regardless how detailed or how exact,

is mere symbology, and can't displace

brain tissue, since it occupies no space.

The student runs a minor risk, perhaps,

of utilizing every last synapse,

and thereby crowding out some mental note

he probably will never want to quote.

But that's the only hazard students met

when autoteaching caused them to forget

inapplicable odds and ends·they knew.

(A few forgot their names, but very few.)

It hardly mattered, when machines that teach
put so much useful knowledge in their reach.

"How come," the poet asked, "it's in this store
of junk you've got no use for any more?"

"That's difficult to answer," said the guide,
"till background information is supplied.
I could explain it to you, word by word;
but isn't such verbosity absurd
when ignorance can be enlighted fast?
Just slip this headset on, for one quick blast."

The Scribe recoiled. "No thanks! I'll learn your lore
the hard way--I'll appreciate it more!"

The human set the two electrodes down.
Perhaps his cowl concealed a pensive frown;

perhaps it didn't. All the apes could see,
despite the lights, was black opacity
that indistinctly filled the hollow space
inside the cowl, which should have been a face.

"You're very foolish," he pronounced, "to fear
the education we could give you here
although I don't suppose there's any need
to hurry things. Since your benighted breed
has lived for many million years without
our aid, you'll last another year, no doubt.
ought to keep in mind that humans, too,
were once afraid of learning something new.
The attitude is prevalent, one finds,
among all species ruled by Lower Minds.
You'll quickly alter attitudes like those
when you're involved in education's throes--
the simplest calculus becomes a pain,
if one assaults it with the naked brain

"You asked me why machines of this design
to obsolescent status we consign.
I'll try to make the matter clear to you
in terms you won't too grossly misconstrue:
Organic neuron circuits, when they're crammed
with much detail, are all too quickly jammed.
In modern times, we've no more need to tax
the human memory with reams of facts.
To your unanswered question, I reply
it's vastly more efficient to rely
on telemetric channels, which connect
my mind to central data banks direct--
for just as tools extend the forearm's length,
computers amplify one's mental strength.
Our forebears weren't as well equipped as we
to use the data banks exclusively;
but we, with these facilities at call,
would need no education now at all.
Perhaps you've noticed, everything I say
is heard from speakers several yards away.

The Central Data System, you'll surmise,
provides my conversation. It replies
to all the questions you see fit to ask.
If asked why _I_ do not assume this task,
the answer is that humans don't demean
themselves with work done better by machine.

"So now you've seen the data problem solved,
but there are several further steps involved
in forging information into fact.
You'll see how later barriers were cracked
with other hardware in this gallery.
It's further down the row. Just follow me."

His Intellectual Achievements

The human floated on. His students walked
behind, all eyes and eardrums as he talked:

"The basic functions of the Mind are three--
Remembrance, Thought, and Creativity.
Recalling information from the past
was now efficient, accurate and fast.
We weren't confined to what we'd heard or seen;
our brains had access to the Whole Machine.
With need for human memory erased,
we found that other. things could be replaced,
beginning with those processes that ought
to be precise and formal--namely, Thought.
One limitation of the mortal brain

is how much data neurons can contain
and simultaneously process. Man
has thirteen billion brain cells, which he can
apply to any problem. That's the sum
and total neuron-count, the maximum;
and most of these are tied up constantly
in passive functions of the Memory,
because, the way the human brain's designed,
these unrelated functions are combined.
Undifferentiated neurons get
too many functions, in a 'random net.'
But that was not the only problem, nor
the one requiring first attention; for
to generalize a Law from facts we know
by brains a lone, was tedious and slow
and prone to Error. Man's organic mind
was not primevally for Thought designed.
It was, instead, a sort of compromise
between too-quick reaction and too-wise.

"When, earlier, I said the human race
has been the toughest foe we've had to face,
I spoke riot only of the civil wars
which modern, cultured humankind deplores;
because our greatest foes were mental quirks
arising from the Lower Mind. This lurks
a quarter inch below the Consciousness,
and is a common source of mental stress
for species who evolve too rapidly
from witlessness to Rationality.
and have to live with Scientific Law
while geared to Nature, red in tooth and claw.
Insinuating instincts-in-disguise,
this atavism makes one rationalize
with specious 'logic,' to support absurd
and preconceived opinions--afterward.
The hardest thing to bring beneath control
was faulty wiring in the human soul,
in which, although the Reason was in charge,
our primitive responses figured large,

and mental discipline must need be used
lest thinking with emotion be confused.
A human's predilection to prefer
believing what he merely wishes were
became a stumbling block as we progressed.
We found that we'd unwittingly repressed
the very knowledge we had hoped to find,
which happened to annoy the Lower Mind
Some 'thinkers' even wanted to invent
tendentious substitute for Measurement.
Such things as 'intuition,' 'common sense,'
and 'shrewd surmise' replaced Intelligence.

"The unworked data we obtained was sound;
but pitfalls on the road to Truth abound.
Inductively, a general truth's attacked
by sifting tons of ill-assorted fact,
until some genius manages to draw
from this morass, a Scientific Law.
Deductive logic had straightforward rules.

Induction, though, was not a game for fools.

One hits or misses, trying to obtain

objective answers from a hybrid brain

which only on occasion can unmesh

its function from the passions of the flesh.

Our aptitude for objectivity

was mingled, almost inextricably,

with ruder talents, suited to a life

of tooth-and-claw survival, lust and strife--

emotional reactions, which detract

from one's pursuit of Scientific Fact.

As you can see, it clearly wouldn't do

to found a theory on a fact or two

when any evidence that doesn't fit

the scheme will instantly demolish it.

That's quite a stringent test for truth to face:

It has to work in every single case.

A searcher after knowledge can't afford

to let a single datum be ignored--

but no one in the world had time enough

to sift such monumental stacks of stuff.
For centuries, new theories rose and fell
as gaps in knowledge gave them time to jell,
and led a generation's scientists
to grope in false--but well established--mists,
while all the crucial information lay
in journals they had 'planned to read, someday.
Besides, too many thinkers tend to choose
which facts to set aside, and which to use,"
according to criteria which fall
a little short of valid, you'll recall.
What use are flawless measurements, if these
are stacked and warped to fit hypotheses
which biased wishful thinkers then defend
--with slanted data--to the very end?

"The absolute necessity was seen
to do constructive thinking by machine.
The reason-function of the human mind
in algebraic symbols was defined,

reducing all complexities of Thought
to AND and OR and NOT and ONE and NOUGHT,
and codifed according to the rules
abstracted from an algebra of Boole's.
Then circuits were designed to synthesize
our own capacity to analyze,
so scientific knowledge had exact
relationship to all objective fact.
Validity for theories finally went
to ninety-nine point nine nine nine percent.
The present aspiration of our race
is knowledge to the hundredth decimal place.
That's quite a grand ambition, we admit,
to set as goals for our ingenious wit;
but one that we can readily obtain
by circumventing defects in the brain.
You must agree, we're coming right along
from times when most of what we 'knew' was wrong.
How close to perfect knowledge we can get
remains to see. We're hardly finished yet!"

He paused beside an ugly, square machine
among the rows of those they walked between.
It bore a marked resemblance to the last
three dozen others they had wandered past.
Perhaps the rows of pilot lamps that graced
its panels were a mite more closely spaced;
but only someone expert at the art
of running them, could tell the things apart.

"This handles data in a larger batch, "
the human told them, "and with more dispatch.
In microseconds, this device could scan
all current information known to Man.
Unerringly it spots one needed fact
in kilo-reams of data, densely packed,
where you could search a thousand years or more--
then miss the very fact you're looking for."

"Let's see it count to five, " the poet said.

The human thumbed a button. "Go ahead."

A screen ignited: 001, 010, 011, 100, 101.

"Great Guns!" the Scribe exclaimed. "The thing's alive!"

The thinker asked, "How much are two and one

and twelve and six and twenty-'leven son?"

The screen lit up: THE PRODUCT OR THE SUM?

BE JUST A LITTLE MORE SPECIFIC, CHUM

"Come on," the human said impatiently.

"We're wasting time. There's quite a bit to see."

His Creative Spirit

"The only thing computers couldn't do"
the guide went on, "Was think of something new.
They couldn't write a program of their own.
They were confined to facts already known.
To put it bluntly, what computers lacked
was Creativity. So we attacked
this problem: Could a circuit be designed
to duplicate those functions of the Mind
by means of which new questions can be posed,
investigations launched, and dead ends closed?
Despite the data they had access to,
we had to tell computers what to do.
They lacked the human Mind's capacity

to look beyond a proved veracity
and seek new questions needing to be solved .

"Fresh insights to psychology evolved
from our attempts to fit hypothesis
to microcircuitry. For instance this:
Ideas, never hitherto combined,
are juxtaposed at random in the mind,
and all our highest mental centers do
is spot those combinations which are New.
Old concepts, thereby made to coincide,
may stem from sources scattered far and wide;
but some New Thing Beneath the Sun will hatch
when two idea-patterns chance to match.

"When once this principle was understood
we saw that cybernetic circuits could
be doing all the 'spadework' humans were
without our tendency to fudge and err.
A human thinker's drawback is, he fails

to check new concepts down to fine details;
so superficial similarities
are apt to cover gross disparities.
When Art, and Science, and Theology
were found to show some rough analogy,
philosophers were prone to think perhaps
their Fundamental Essence overlaps;
and, in their hopes some brave new Truth to learn,
ignored conflicting facts at every turn.
A cybernetic brain, however, takes
the time to check for possible mistakes
enabling it more promptly to reject
conclusions which are subtly incorrect.

"This was, like everything we'd ever done,
no sooner formulated than begun;
and, once begun, effectively complete--
an epoch-making engineering feat.
The problem was in Fortran Code expressed;
magnetic tapes were furled, and buttons pressed.

And, lo, we had replaced our minds. We've seen
new questions formulated by machine,
reduced to algebra, and then resolved
without a human scientist involved."

He showed the apes another square machine:_
exactly like the others they had seen.
They watched the rows of neon lightbulbs blink.

"Now this, .. the human said, "Can really think."
He opened up a door of painted tin.
The tribesmen squatted down to peer within,
where they beheld wondrous, Gordian maze
of wires that ran a hundred different ways.
It didn't really tell them too much more
about computers than they knew before.

"By analyzing our imaginations
into random thought-associations
human engineers could automate

the 'purely human' power to create.

And <u>this</u> machine is all machine designed--

a real triumph for the Human Mind!

"You've seen enough, for now of all the vast

accomplishments that glorified our past.

There's something more contemporary,

we believe the two of you would like to see."

He led them to a wall. When they were there

a door-shaped hole dissolved to empty air.

The human ushered both the tribesmen through.

The door behind them disappeared from view.

They stood there, blinking, in the glaring sun,

not far from where their tour had first begun.

Their eyes adjusted, and they looked around

at charcoal-colored walls.

The thinker frowned.

"You should do something to improve this view,"

he said, "like painting buildings red or blue.

We apes have always noticed, multihued
surroundings brighten up a person's mood."

The human laughed. "Illusion pure!" said he;
"a matter of photonic frequency.
You've heard it said that roses smell the same
by any other arbitrary name.
A wavelength has less meaning than a word,
so why should certain colors be preferred ? "

While they were talking, they had crossed the street.
The tourists watched their human guide repeat
the scientific way to make a door
where there were only solid bricks before.
This time the Scribe, from curiosity,
hung back a moment, surreptitiously,
to see if he could make a door appear
in solid-looking walls, by drawing near.

He felt much better when he found he could.

The wall, and not the Creature in the hood,
had somehow been the automatic cause
of this apparent lapse of natural laws.
The mechanism wasn't his concern.
A pragmatist, he was content to learn
to make things happen--it was mankind's quirk
to have to know how mechanisms work.

His curiosity allayed, he went
to join the group, and hear the argument
the human and the Sage had under way.
As he approached, he heard the human say:

"True Science rose from our proclivity
for non-emotive Objectivity.
We'd compromise it, if we changed the hue
of our surroundings to your red and blue.
And even if an optimistic tinge
could on the human intellect impinge,
what reason could a thinking person find

to tolerate disruption of his mind
by letting colors, smells and sounds intrude
upon his thoughts, by altering his mood?"

"But that's ridiculous! " the Sage exclaimed.
"If that's the frame of mind to which you aimed,
you'd reach a stage where you were never stirred
by any sight you saw, or sound you heard.

"Precisely," said the human. "We observed
that in pursuit of Fact, emotions served
no purpose; and in fact effectively
prevented reasoning objectively.
It's only when a scientist will strive
to ban involvement, that researches thrive."

"I think you're spoofing us," the poet said.
"Without involvement, you'd be better dead.
You're <u>sure</u> you've got no passions now at all?"

"They're vanquished "said the guide, "beyond recall.
Abandoning emotional response
facilitates objective nonchalance.
We found the means completely to submerge
and neutralize the procreative urge--"

The Scribe exclaimed, "What's <u>that</u> improvement's use;
and how the hell to humans reproduce?"

"What purpose would it serve?" the guide replied.
"It's been so long since any human died
it's quite unnecessary to replace
our generations, to maintain the race.
We just install a few replacement parts
--electric or hydraulic brains and hearts--
tune up his hormones and his glycogen
and get the patient on his feet again.
To let trained research workers be replaced
by callow youths would be a tragic waste."

The thinker scratched his head. "I see your point,
but somewhere your conclusion's out of joint.
There's something there that doesn't <u>fit</u> too well--
but let me think it over for a spell."

The Scribe spoke up: "I've got a question; too--
just what the hell does <u>this</u> contraption do?"

He'd paused before a maze of wheels and cogs
and pawls and rocker arms and ratchet dogs,
and compound pulleys linked with belts and springs,
with several banks of blinking lights and things
where tubes emitted their magenta glow,
and meter needles flickered to and fro.

"Ah, yes!" the human said; "perhaps the one
most all-important project we've begun.
This pilot model (which will later be
condensed, I'm sure) is what we've come to see.
The data and technology derived

from every project mankind has contrived
have now converged, as this device evolved,
to solve all questions hitherto unsolved.
If all discoveries we've ever made,
and all inventions, in the long parade
that stretches from the prehistoric past
can culminate in one device at last
--one monumental tribute to the wit
and spirit of mankind--then-this-is it!
It culminates our most ambitious goal:

mechanical replacement for the soul."

HIS DESTINY

The apes could think nothing much to add.
They both looked deeply thoughtful, wise and sad.
The Sage sat down upon a motor case
with cogitation written on his face.
The lull continued for a moment more.
The human was about to speak, before
the poet cut him off:
"No; hold it, Mac!
Our Sage's mind is strictly single track,
so let's not interrupt his train of thought
until this fleeting spark of wisdom's caught.
The final fruits of his analysis ·
are nothing you can well afford to miss.
When his cerebral neurons start to click,

he'll leave computers looking pretty sick;
so brace yourself for something really hot,
'cause he's the wisest thinker Eden's got!"

The lull retained the floor, and no one spoke
until the Sage himself the silence broke:
"Some deep and careful thought," he said, "would seem
to show the human race is off the beam.
You think you've really got it over apes,
despite our race's more efficient shapes,
because-you say-of all the things you know
and wonders you've accomplished. Even so,
you need machines for everything you do.
My question is: How much of you is _you_?
You've said a half a dozen times, your goal
was bringing Nature under _your_ control;
but (here's the point I want to put across)
just who, around this town of yours, is boss?
You've got computer systems here and there
that tell you what to do, and how, and where. -

They tell you what to think, and what to say.
Your lives are run some dumb computer's way!"

"You apes," the human scoffed, "are utter fools!
Machines are nothing more than useful tools!
However competent computers seem,
they're our creations. Man remains supreme!"

The thinker waved the argument aside
"You missed my point completely, " he replied.
"Machines like these would very soon replace
all reason for existence, for your race.
I'm mortal; but I'll live and die an ape,
and not some electronic super-shape
that talks in four-bit words, and wears a cowl,
and somehow isn't either fish nor fowl.
You think a human's smarter than an ape,
but all your brains are on magnetic tape!
What other beasts, except for you, rely
on circuits to subtract and multiply?

An ape believes in anything he wants
by disregarding skeptics' lies and taunts;
but humans need computers to reject
conclusions it decides are incorrect.
Whatever brains I have are part of me:
they don't depend on microcircuitry.
If numbers over twenty flake me out,
then that's a talent I can do without;
but pushing buttons on some damned machine
with blinking nixie tubes, like what we've seen
can't make you any smarter than you are!

"I'm tougher than a human is, by far;
and all those guns and neutron bombs you've got
just go to show you what a human's <u>not</u>!
I guess you've maybe heard it said a fight
won't settle which contestant's in the right;
but, even so, it ought to prove who's <u>strong</u>-
- and humans rig it so the outcome's wrong!
The <u>only</u> beasts, from ants to pachyderms

that spend their lives in deathly fear of germs
are human beings. Others only strive
to lick their wounds and somehow stay alive.
Why don't you be what you're supposed to be
instead of mimicking machinery?
You don't find dogs that try to act like cats-
or, if you do, you think they're slightly bats.
A cactus doesn't act like turnip greens;
so why should humans try to be machines?
If just a human's all I was, by damn,
I'd try to learn to live with what I am!"

"Whatever your impressions," said the guide,
"we've lived, while less 'unnatural' species died.
Each creature finds its adaptations, which
define its normal ecologic niche;
and human adaptations, we maintain,
are cumulative products of the brain.
To say our 'artificial' human way
is 'anti-natural' is as if to say

that birds should not build nests, or that the moles
are wrong in digging artificial holes.
The very fact that Man is still around
has proven that his adaptation's sound."

The thinker scowled, and firmly shook his head.
"You've tried to change the subject, now" he said.
"The battle for survival must be won,
but still, the <u>fight itself</u> is half the fun.
I know the only reason <u>we're</u> alive
(and I mean <u>why</u>, not <u>how</u>) is to survive.
We hunt for meat because we've got to eat
to have the strength we need to hunt for meat:
a vicious circle, sure; but can't you see
that that's the way your life's <u>supposed</u> to be?
One Universal Law we apes have found
is, Nature keeps revolving, 'round and 'round.
The sun comes up and then goes down, and then
revolves around the world to rise again.
New generations come, and old ones go,

and tell the same old tales of grief and woe.
I teach my cubs the lore I know, so they
can pass it on to cubs of theirs, someday,
and so that maybe when they come of age
they'll qualify to take my place as Sage.
But that's the way of Ape, and not of Man-
you store your racial wisdom in a can
of ferrous doughnuts! Any fool can pour
a silly gadget full of coded lore
and pull prefabricated answers out,
pre-checked, without a decimal-place of doubt!
But who takes credit for a problem solved
unless a human brain has been involved?
When new, machine-discovered truths arise,
do wires and buttons win the Nobel Prize?
When rockets scorch a foreign landscape brown,
do neutrons come to claim a laurel crown?
Are you alivc, if all you'll ever be
depends on how you use machinery-
or is it time to relegate yourselves

to antiquarian museum shelves?
Just everything you humans used to do
has been fobbed off on your computer crew!
That's why I said your logic's out of joint:
You humans have completely missed the point.
I'm starting now to wonder whether you
can do as much as apes have failed to do.
A human's tough because he packs a rod;
but let me ask you this: Have you found God?"

The human gave a snort of sheer contempt.
"From such illusions we've been long exempt!
You can't define-you merely verbalize-
this 'god,' which by its very nature lies
beyond the reach of rationality
and thus possesses no reality.
It's just a word, applied to 'that which' lies
beyond your observation or surmise.
Such quantities are not in our domain
of concrete measurement. It will remain

unknown, unanalyzed and unobserved.
The only use such concepts every served
has been as fuel for the smoky haze
that fogged our minds, in less pragmatic days.
On such a theme, I'll say no more than this:
We have no need for that hypothesis!"

"Aha !" the Scribe broke in. "I get the gist!
You're just another stupid atheist!
We don't need you as guide, to search this place!
You're fired! ... Before you go, let's see your face--
this peek-a-boo's been carried far enough!"

With that, he seized the human by the scruff.
Both tribesmen pounced upon the guide to tear
his robe away, and

there was no one there.

The robe and cowl fell limply to the floor.
There wasn't any human, any more.

The tribesmen gripped their clubs and edged away.
The poet's shaky voice was heard to say,

"We sure enjoyed your hospitality,
but now, heh-heh, we'll change locality.
This little visit's been a lot of fun,
old friend; but now I guess we'd better run"

The city's portal still stood open wide.
Had anyone been there, he might have spied
two blurs of motion, charging up the street,
or heard the fast, staccato thud of feet.
A pair of apes, with all their fur on end,
came streaking out the gate, around the bend,
and caromed madly down the spiral ramp,
then off across the plain, toward Eden's camp.
They dwindled rapidly, till they were just
the distant ends of slender clouds of dust.

What ape can be condemned as cowardly

who never shrank from perils he could see?

Our spies had done their job, and done it well,

returning safe with quite a tale to tell.

Would any beast of flesh and blood have faced

the threat of being totally

replaced?

ABOUT THE AUTHOR

Jon P. Gunn wrote The Apes of Eden over a period of many, many years. He read Spenser, Chaucer, Dante and Cervantes. He died while still intermittently adding chapters. Many oddball philosophies, from solopsism to deism are explored and mocked. Allusions to a broad spectrum of myths and canons are made.

Jon never graduated from college even though he had twice the number of hours to graduate. He was too busy reading the great works of literature to bother. He shared the Apes of Eden with a good friend. It is that man and his friend, Rick Lakin who are bringing you Jon's work. We think it's very good. We hope you do too.

Made in the USA
Monee, IL
04 November 2026

67377505R00075